L7

Aa. H. KAMPP

AN AGRICULTURAL
GEOGRAPHY OF DENMARK

AKADÉMIAI KIADÓ · BUDAPEST 1975

Translation revised by

PAUL A. COMPTON

The Queen's University of Belfast
Northern Ireland

ISBN 963 05 0673 4

CONTENTS

PREFACE

The only natural resource of importance in Denmark is the soil. Although it is not particularly fertile, the prosperity of the country was built up during the 19th century by means of agricultural exports first and foremost to Great Britain.

In Denmark the co-operative movement was founded in 1866 and during the last decades of the 19th and the first part of the 20th century it expanded to such an extent that the greater part of agricultural activity was based on co-operative organisations, a procedure which furthered efficiency enabling Danish farmers to produce top-quality food products.

Today industrial goods have taken over the lead accounting for 69 per cent of exports by value (1970). But still about two thirds of total agricultural produce is being exported. Although Denmark is a very small country its exports of livestock products are among the largest in the world.

Hence it may seem natural that a book dealing with Danish agriculture should be one of the first in this series, and the author felt it a great pleasure and honour when the Editor-in-Chief GY. ENYEDI invited him to prepare this monograph.

The author would like to express his appreciation to Professor C. THOMSEN, The Royal Agricultural University of Denmark, for his critical review and comments, to K. E. FRANDSEN, University of Köbenhavn, who has reviewed and partly revised the historical section entitled "Open-field system and enclosure movement", and to the Central Co-operative Committee (Andelsudvalget) who provided him with information about the present situation of the co-operatives.

Valby, March 1974   *Aa. H. Kampp*

INTRODUCTION

Denmark consists of the peninsula of Jylland plus an archipelago made up of more than 500 islands, of which about 100 are inhabited (*Fig. 1*). It comprises many different forms of landscape and a mosaic of different soil types from the desolate granite regions of North Bornholm and the dune landscape in West Jylland to the highly productive farm land on the morainic deposits which form the greater part of the Danish landscape. *Figure 2* shows the general geological arrangement, although essential differences may exist within a given holding, indeed, even within a single field.

The climate varies little from south to north or from west to east, the mean maximum temperature ranging from 15° to 16·5°, the mean minimum from minus 1° to plus 0·5° centigrade, and the mean annual precipitation from 500 to 750 mm. Westerly winds predominate and are, of course, strongest towards the west (*Photo 1*).

Photo 1. West-wind swept hawthorn in a windbreak, Vendsyssel (by Kampp)

Fig. 1. Denmark, outline map

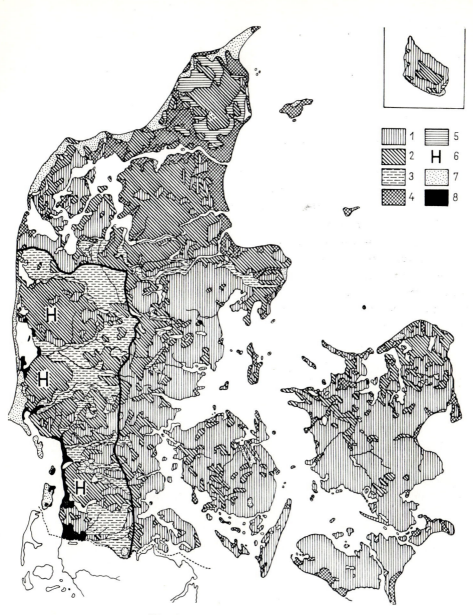

Fig. 2. Denmark, surface geology

1 = moraine, mainly clay; 2 = moraine, mainly sand; 3 = outwash plain; 4 =
litorina deposits; 5 = yoldia deposits; 6 = hill islands; 7 = dunes; 8 = marsh.
The unbroken line represents the limit of the main advance of the ice during the last
glacial period (Würm)

A. HISTORY

OPEN-FIELD SYSTEM AND ENCLOSURE MOVEMENT

During the Middle Ages the land gradually came into the possession of the nobility or the Crown, and by the middle of the 17th century most of the farmers had become tenants. The greater part of the land, however, continued to be cultivated by tenant-farmers living in the villages. The Danish feudal system can be described as a *"Grundherrschaft"*, the landowner relying on income in money or kind (*landgilde*) and some work performed by the tenants, rather than a *"Gutsherrschaft"*, in which he would draw profits from the sale of serf-produced goods as in Eastern Europe. The land operated directly from the manor was relatively small compared with the peasant-operated land.

In the open-field system the land was individually farmed, but the tenants had to submit to joint decisions; individual initiative was impossible (Kampp and Frandsen 1967).

Up to the year 1800, the open-field system in its different forms was in use in most of Denmark, and only minor parts of western and northern Jylland comprised areas with small, enclosed single-farms. In Sjælland and the eastern parts of Fyn and Jylland the three-field system prevailed. In other parts of Jylland various other systems were in use.

A radical change took place after the Enclosure Act of 1781 when it was laid down that every holder had the right to exchange his cultivated land as well as his part of the meadows, commons, and moors for a compact holding even if the rest of the inhabitants of the village were against re-allotment (*Fig. 3*).

Most of the farms were adequately consolidated, which involved advantages as regards labour, transport, and crop rotation, thus advancing effectiveness and competitiveness. Expenditure on surveying and mapping was shared and the map had to be approved by all the peasants of the village (*Fig. 4*).

A peasant was to have the same amount of assessed value of land after re-allotment as before. The outline of a property had to be as appropriate as possible. If the length was more than 4 times the width or if the buildings were more than 1500 *alen* (about 900 m) from the most distant part of the

12

farm, an outlying farm was to be established with economic support for erecting new buildings. Almost simultaneously the drainage of numerous swamps was begun, whereby the tillable land area was expanded considerably (Kampp and Frandsen 1967).

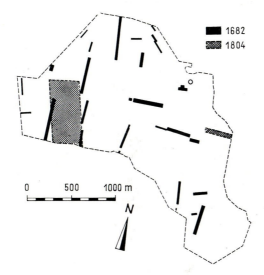

Fig. 3. A single farm in a village before and after consolidation. The small circle shows the location of the buildings of the village during open-field period (after Kampp and Frandsen 1967)

An act of 1794 concerning fences and the right to prevent trespass made every farmer responsible for the cultivation of his land, thus enabling him to introduce new crops and to develop better livestock breeds, which was impossible while stock grazed in common. Better agricultural implements, drainage, and marling created far better conditions for crops. A series of technological improvements in the following decades coincided with operational changes. A transition from subsistence output to market production gradually took place. Concurrently with this development the village pattern changed.

During the course of the nineteenth century an increasing number of farms became privately owned as the augmented yield of the land created better economic conditions for farmers, and towards the end of the century the greater part of Danish farmers were freeholders.

During the last decades of the nineteenth century Danish grain exports declined on account of reduced market possibilities, and as agriculture was the only significant national occupation there was no possibility for effective economic support schemes for agriculture as in several other European countries. The farmers shouldered the consequences and shifted towards stock-breeding. Animal products became the main branch of exports

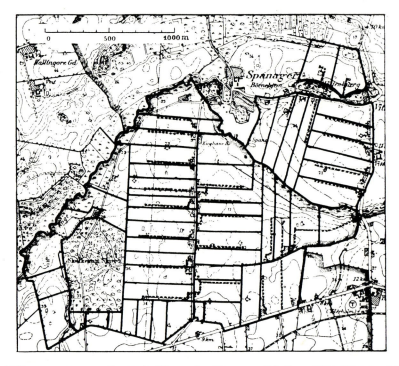

Fig. 4. The same area as shown in Figure 3, after consolidation, 1804 (Kampp 1965b)

particularly to England, where agriculture was not able to meet the demands of a rapidly growing industrial population. Simultaneously the co-operative movement for sale and purchase progressed in Denmark, and most processing works such as dairies and slaughterhouses, were made co-operative so that the farmers themselves controlled the goods from producer to consumer, the principle of "vertical integration".

THE PARCELLING-OUT PERIOD

As far back as about the year 1800, especially in the decades following the enclosure period, owners of manors began erecting thousands of small-holdings in an effort to secure labour, and actually the number of holdings

14

grew considerably faster during the nineteenth than during the twentieth century (Kampp 1959). During the last decades of the nineteenth century a large-scale emigration to urban areas and the New World created a dearth of manpower in rural districts, at the same time as the expansion of animal husbandry was creating more work than the former grain production. In order to counteract this development the State took the initiative to create better means of subsistence for farmhands, and in the year 1899 an Act of Parliament introduced State-subsidised small-holdings for rural labourers. The size was limited to 1—4 hectares according to soil quality and State loans were granted.

As a result of changing views regarding unjust social conditions claims were made in the following years to make the holdings large enough to support a family. Ten years later, in 1909, the size limits were raised and at certain intervals thereafter the existing limits were further extended. Small-holdings already established were once more granted loans from the State to purchase additional allotments. The request to have the holdings enlarged was a manifestation of the need for economic and social independence.

In 1919 new laws were passed with the same intention, namely to create a large number of small-holdings. This time, however, the land was State-owned and could be rented by the small-holders. The available land for this purpose was glebe and other areas owned by the State.

An important provision of one of the Acts laid down that owners of entailed estates should make over to the State one third of their area together with a substantial sum of money, in return for which they obtained proprietary rights to the estate (*Fig. 5*).

Between 1900 and 1950, many of the 30,000 State small-holdings evolved to form characteristic modern villages — comprising an element prominent not only in the settlement geography, but also in the topography of the country with the division of large fields into smaller plots, lending new features to the entire cultural environment (*Photo 2*).

The State-subsidised holdings meant social reorganisation and fitted well into the desired economic development where the intensively operated farm was advantageous. Animal husbandry was comparatively important and accordingly production per area unit was high.

In many places the last traces of the more delicate structural aspects of the natural environment disappeared completely as a result of intensive cultivation. Microtopographical features especially were transformed by the removal of ditches and dikes, knolls, small bogs, field boundaries and other details so insignificant that they were often not included in the list of signs on the Ordnance Survey map. Fields became more regular in shape,

15

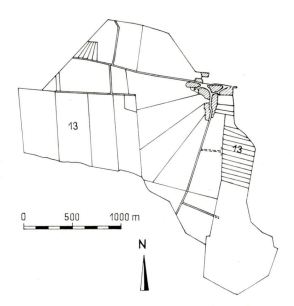

Fig. 5. The parcelling-out of Spanager Manor (Själland) during the conversion of entailed estates into free simple holdings in 1923, drawn on the Ordnance Survey map (Kampp 1972)

except when prevented by the profile of the terrain, larger water-courses or roads, woods and parish-boundaries. In the old cultural landscape roads were secondary features conforming to the already existing pattern of habitation. In modern time, roads link towns and villages, and increased mechanisation has done away with the necessity for placing the farm buildings in the centre of the fields.

THE MANOR HOUSES

Manor houses have been a characteristic of the environment since the Middle Ages. Today they are monuments due to the architectural styles of the past. Vast fields, usually fenced in by hedgerows or stonedikes indicate the presence of manor houses. They have left their marks, pleasantly and permanently on the scenery. Most of the manors were placed in the

Photo 2. Aerial photo, showing the estates of Gaunö, Själland and Lindersvold at the same scale, split up into small-holdings in 1922 (by Geodät. Inst., 1939)

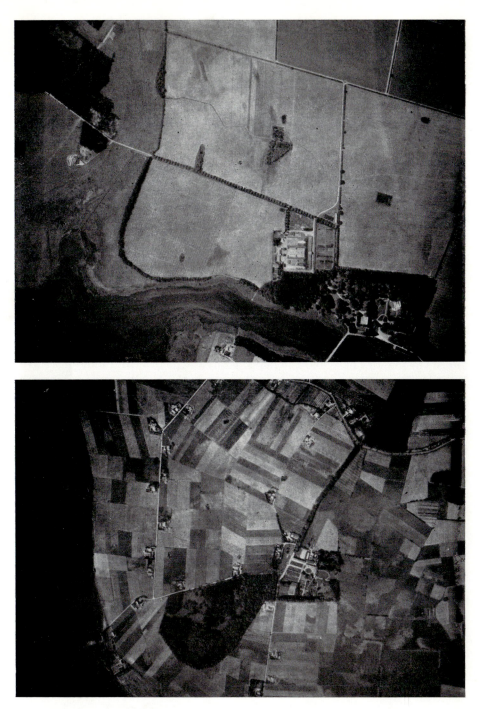

Photo 3. Silo and stacker in front of a manor house, South Själland (by Kampp)

fertile parts of the country, where the predominantly clayey morainic hills exhibit a variety of scenic beauty. Not infrequently, the manors were erected in the very centre of such beautiful landscapes, surrounded by natural parks or artificial "green architecture", from which long drives lead past extensive fields, and where stands or single trees sometimes bridge the distance between the manor house and the fields. Many of the beautiful buildings have long since been deserted in their neglected gardens, while others, protected by law are being preserved and kept in good repair. As a historic element in the progress of land-cultivation the idyll is nowadays being disturbed by features of modern, large-scale industrialised farming, adding gigantic silos and functionally designed barns to the original buildings (*Photo 3*).

B. THE CO-OPERATIVE MOVEMENT

The close human relationship between the farmers of the village during the open-field period was to a certain extent broken when a great many of them moved to outlying farms during the first part of the 19th century. To a certain extent, the feeling of having things in common was revived with the spread of the co-operative movement.

The first Danish Consumer's Co-operative Society, founded in 1866 and based on the Rochdale principles, was very soon followed by many other co-operative consumer's stores. In 1882 a very decisive step was taken with the foundation of the first co-operative dairy. And as early as about 1900 the most important branches of the Danish Co-operative Movement had been founded, e.g. co-operative slaughterhouses, egg-exports, butter exports, imports of feeding-stuffs and fertilisers, etc.

In the second half of the 19th century there was a tendency in economic life toward large-scale enterprises and concentration, showing the typical trend towards the weakening of the personal relationship between the different stages of production and marketing. Protection against the unfavourable results of this trend was the first and foremost object of the co-operative movement and was in most cases the immediate reason for their establishment.

The rapid growth of the movement was furthered by educational conditions in Denmark. Firstly, compulsory education dates back to 1814, so that for several generations there had been no problem of illiteracy; secondly the Danish Folk High School, a very special form of adult education, started in 1844. The 5—6-month courses that they offered became much frequented, in the 19th century especially by farmers' sons at their own initiative, at their own expense, and for the sake of education without any expectation of a degree or material return. They went back to their daily work not only with a religious and nationalistic outlook but also highly conscious of the value of democracy in political affairs as well as in narrower relations. They were thus ready to accept and work for the ideas of the co-operative movement.

The direct causes of the founding of agricultural co-operatives were certain radical changes in marketing conditions during the last quarter of the

19th century. Greatly increased production and marketing, mainly the export of animal products was achieved largely through the new co-operative societies. Between these and Danish agriculture as a whole an interaction started of the greatest importance for both.

The basic rules for all Danish co-operative societies were:

(1) that profits should be divided among members according to their production or purchase,

(2) that the members themselves should elect their committee on the principle of "one man one vote" irrespective of the size of his production or purchase, and

(3) that membership should always be open to new members from the region covered by the association.

This democratic form of government and the equal payment for products delivered to them furthered the possibilities of minor farms and small-holdings. Without sacrificing their individual liberty, independent farmers have carried co-operation further than in other countries (except perhaps Iceland). It plays a very prominent part in Danish economic life, especially in the branches where there is a lively trade with other countries and is thus important for farming in all its aspects. In the two most important lines, namely pigs and milk, by far the majority is handled by the co-operatives. More than 90 per cent of all Danish farmers are usually members of several co-operative associations for processing, sale, purchase, credit and insurance, and other services.

Up to the 1960s the typical co-operative body was the local society with open membership, obligations to deliver or purchase, personal joint liability and democratic leadership. But since then the structure of Danish co-operative organisations has been undergoing changes the main feature of which is a move towards concentration.

Many of the small societies are now being amalgamated into larger units, while some are being dissolved and the members spread among neighbouring societies. This trend is of course aiming at the establishment of co-operatives of a proper size, i.e. large enough to operate the most up-to-date equipment.

The unlimited joint liability which was usually found in small societies has often been changed into limited liability, when the larger unit is created. The former direct election of board members has been substituted by indirect elections through an assembly of delegates, but even under the new conditions the aim is to have as close a member influence as possible. In some sectors the final goal is one country-wide society. This status has already been achieved in some instances, while in others it is being created.

There are sectors too where the aim is to have regional societies as the first goal and where it is uncertain whether they will evolve into national bodies.

Consequently a survey of Danish co-operatives today will exhibit some of the old and some of the new structure. Some have already made the change, some are well on the path and others are only just beginning the transformation.

C. MODERN AGRICULTURAL PRODUCTION

DISTRIBUTION OF HOLDINGS AND FARMLAND

Towards the middle of the 20th century the parcelling-out movement was superseeded by an opposite trend favouring larger farms which have better possibilities of adapting to modern methods of farming through increased mechanisation and to changing post-war market conditions. Parcelling-out gradually ceased, amalgamation was finally legalised in 1962, and now the number of farms is declining with accelerating speed. Of the 44 holdings of Spanager (see Fig. 5) in 1969, 26 were being run by 13 farmers in addition to which three other farms had been rented by farmers from outside the area (Kampp 1972).

Labour is today being attracted by the more permanent and better paid jobs in manufacturing industry. Consequently those who remain farmers have the possibility of making larger profits.

Since 1960 the number of holdings has fallen by one third, although in Jylland only by a quarter. Two thirds of the farms cover an area of 10—50 hectares (*Fig. 6*). This development has changed the distribution of land but not the pattern of settlement.

Owner-operated farms have been dominant in Denmark for more than 150 years, and today most Danish farmers are freeholders; only 3·8 per cent of farms are rented. Tenant farmers, however, enjoy great independence in making decisions, several farmers have begun to realise that the right of use of the land is more important than ownership, and it has become more common for a landowner to rent the land of a neighbour who then only uses the buildings as a residence. In 1964 joint operations were twice as common in the Islands as in Jylland.

Farms and holdings are fairly evenly distributed throughout the country. As a consequence of soil fertility, holdings of less than 10 hectares are found mainly in the eastern part of the country except in Lolland and Falster, where manors have survived to a larger extent than in the rest of the country. Farms with more than 10 hectares are found mainly in West and South Jylland, and over 50 hectares in South Jylland and Lolland. This can to a certain degree be seen in *Figure 7*, showing the average size of holdings over 0·5 hectare.

22

Up to the middle of the 1960s land was being reclaimed from moors, lakes and the sea (*Photo 4*), often with the help of State-subsidies, but because of the cost this has ceased, and only marshland on the west coast is still being reclaimed, partly naturally, and partly by human intervention.

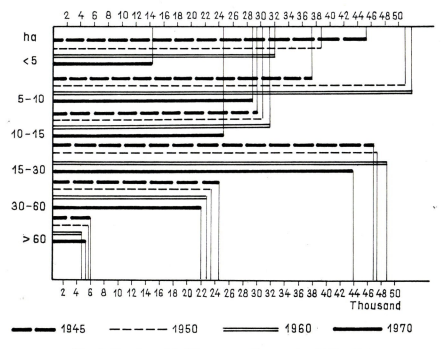

Fig. 6. Number of holdings according to size 1945–1970

The reduction of farmland since 1961 (*Table I*) has to a large extent occurred at the expense of the most fertile parts of the rotation area; the largest decrease has taken place in the Islands and eastern Jylland, but a parallel development is already noticeable towards the west.

The cereal areas decreased between 1938 and 1950, but the reduced labour force, mechanisation, and effective pesticides and herbicides have since then brought about a strong upsurge (*Fig. 8*).

Additionally the possibilities of growing other cash crops are very limited, while the relatively low prices of animal products up to 1973 have brought about a reduction in feed crops.

The accession to the EEC in 1973 means altered economic conditions for Danish agriculture. For the first time in many years 1973 showed a de-

23

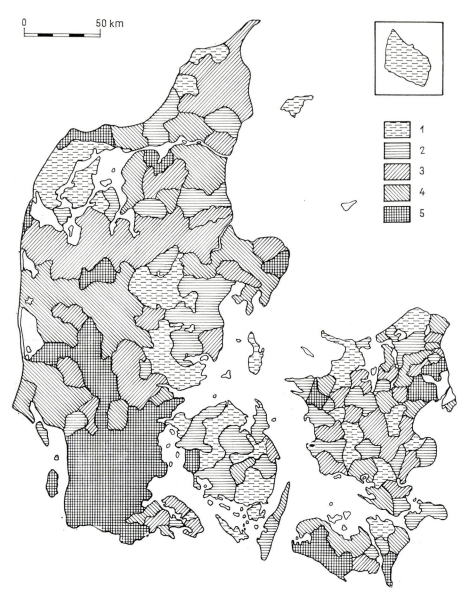

Fig. 7. Average size of holdings, 1971

1 = ≦20; 2 = 20·1—22·3; 3 = 22·4—25·2; 4 = 25·3—28·5; 5 = ≧28·6 hectares

Photo 4. Reclaimed land on old seabottom, North West Själland. In the background buildings belonging to small-holdings from the parcelling-out period of the 1920s (by Kampp)

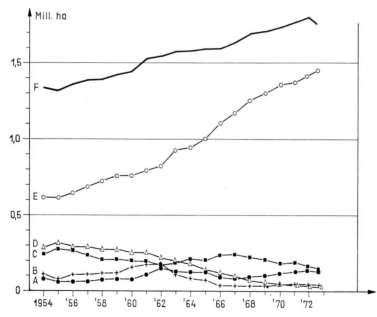

Fig. 8. The development of the cereal area, 1954—1973

A = wheat; B = rye; C = oats; D = dredge; E = barley; F = total

crease in the total cereal area, roughage increasing at the expense of cereals due to better prices for animal products (Thomsen 1974).

Table I

Total area of Denmark	4,306,900 hectares		
	1940	1960	1971
Number of farms	205,000	196,076	135,600
Agricultural area, hectares	3,218,000	3,094,000	2,915,000
Rotation area, hectares	2,657,000	2,752,000	2,626,000
Areas outside rotation, hectares	561,000	343,000	289,000
As a percentage of agricultural area	1950	1960	1971
Cereals and pulse	42	47	60
Fodder roots	13	13	6
Grass and green fodder	36	32	25
Barley as a percentage of cereal area	38	58	78

DISTRIBUTION OF CROPS

The various cereal crops are fairly evenly distributed over the country, Spring barley now comprises 51 per cent of the rotation area, but a remarkable change in land use has taken place during the last few decades. Attention in particular should be drawn to the fact that the barley area has almost doubled since 1960 and in 1973 accounted for more than 82 per cent of the cereal acreage (Thomsen 1974). At the beginning of this century barley was only grown in the eastern part of Denmark, but it has gradually spread at the expense of other types of cereal to the western part as fertilising techniques gained a foothold and new better yielding varieties, even in poorer soils, were developed (*Fig. 9*). A supplementary explanation is the general trend towards larger numbers of pigs.

From a climatic-edaphic point of view Denmark is in a fortunate position in that in general there is plenty of rainfall on the light soil. Also soil improvement plays an important role. Barley will not tolerate acid soils but marling has diminished such areas. Indeed, tests on light soil fields have

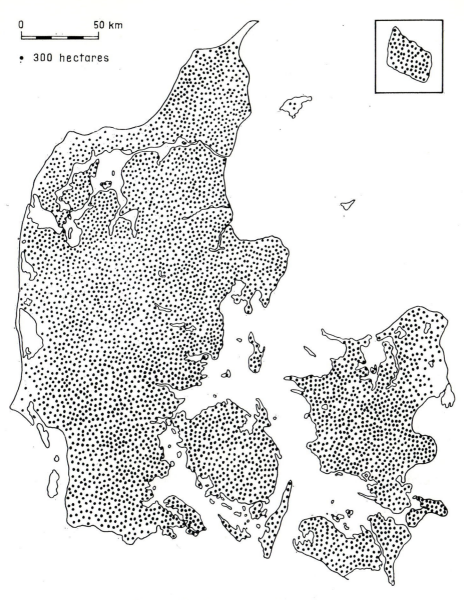

Fig. 9. Distribution of barley, 1971

27

shown that even here barley produces higher yields than oats, which has resulted in a considerable increase in the acreage of barley area on light soils.

The growing of winter barley had to be prohibited over the 5-year period 1968—1973 because it is the host of mildew, yellow rust, and brown rust, which diminishes the yield of spring barley. All other cereals have decreased in area. Until 1938 oats was the most important cereal, comprising the largest area, but since then it has gradually decreased, among other things because of the decline in number of horses (*Fig. 10*, and *Photo 5*).

At the beginning of this century rye was evenly distributed over the country except in Lolland, Falster and Mön. Then it was limited to the light, sandy soil of West Jylland, and now there are only small areas left, mainly in South Jylland (*Photo 6*).

Wheat has always, as at present, its centre of gravity on the fertile soils of the eastern part of the country, where the climate is also warmer (*Fig. 11*). The previously accepted explanation of the preference for barley as against wheat in Lolland and Falster, i.e. that barley was more stiff-strawed, is no longer valid because equally stiff-strawed varieties of wheat have been introduced. Tradition of course, is also an important factor.

Similarly to oats, rye, and wheat, dredge has also lost ground, following a large increase from the beginning of the century up to about 1940. One reason is the introduction of the combine. It is now virtually a Jylland crop.

Even though very short strawed varieties of cereals are used, there is still a surplus of straw; after the harvest it is burned in the fields. Of course, the ashes give plant nutrients to the soil, but no humus, and the strong heat caused by the fire kills part of the micro-organismus.

Pulse is grown especially on the Islands and in the eastern part of Jylland.

Fodder beet-root areas have increased to a certain extent, but the swedes acreage has fallen back because they yield less solids (*Fig. 12*). Root crops, especially potatoes, are mainly grown in Jylland (*Figs 13, 14*, and *Photo 7*).

Sugar beet for the factory has since the 1870s been grown in Lolland Falster, Mön, South and West Själland, and Middle and West Fyn (*Fig. 15*). From 1872 onwards sugar refineries were built in the most fertile parts of the country, and for this reason sugar-beet is still exclusively found in such areas despite the fact that by means of fertilisers the yield would be the same in the more sandy parts towards the west where lifting would be easier (*Photo 8*).

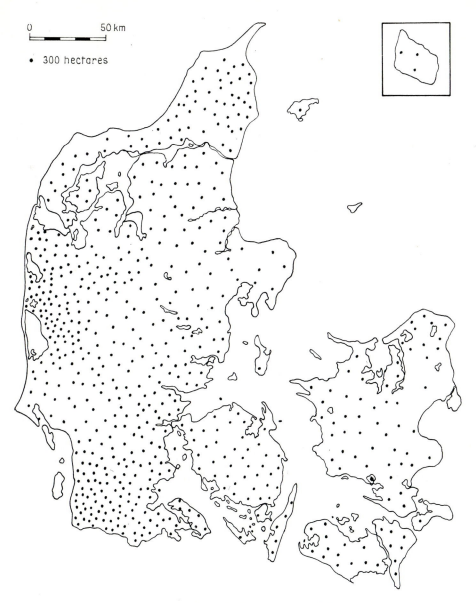

Fig. 10. Distribution of oats, 1971

Photo 5. Characteristic features of the undulating glacial landscapes of Denmark: farms spread over the cultivated land by the exchange of strip holdings about the year 1800. In the foreground an oat field. Mön (by Kampp)

Photo 6. Rye field, windbreak and farm buildings on West Jylland outwash plain (by Kampp)

30

Fig. 11. Distribution of wheat, 1971

31

The six sugar refineries still contract every year with the farmers as to the area to be sown with sugar beet seed. Until the 1930s sugar production was in the hands of private enterprises, but during the period of unemployment the Government intervened and continued to fix the price of sugar up to 1973 when Denmark joined the EEC. Since the beginning of the 1960s, sugar exports have been built up especially to Norway.

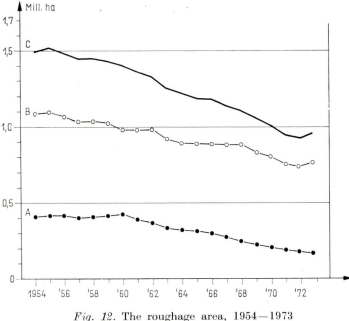

Fig. 12. The roughage area, 1954—1973
A = fodder roots; B = grass and green fodder; C = total

One might wonder why the beet growing areas are not more extensive. In the first place, beet like any other cultivated plant must enter the system of crop rotation, and it has proved disadvantageous to grow beet in the same field at intervals of less than 5—7 years if attacks by parasites are to be avoided. In the second place beet is extremely labour intensive, and there is a limit to the amount of work which a single farmer can manage. Smaller holdings, however, have traditionally been more labour intensive per hectare, and for this reason they have carried comparatively larger areas of beet for fodder. In recent years, however, new types of beet seed combined with herbicide spraying are reducing the amount of manual work in the field.

Fig. 13. Distribution of root crops, 1971

Fig. 14. Distribution of potatoes, 1971

Photo 7. Potato field in springtime behind a windbreak on an outwash plain. A small old-moraine hill island in the backgroud. West Jylland (by Kampp)

Apart from sugar beet, the basis for root crops is stock-breeding. Roots have a positive influence on the yield of cereals and offer the possibility of rotation between deep-rooted broad-leaved crops and short-rooted cereals. Moreover, root crops improve the structure and increase the humus content of the soil in connection with liming and drainage. Another advantage is that weed control is easier and more effective than in cereal fields.

Progress, of course, has not been equally rapid everywhere. Unfortunately, no experimental material is available that enables one to define the importance of soil improvement in terms of increased yield. Manuring was begun on the better quality soil, likewise the distribution of new strains. In the long run, relatively greater improvements in quality have been achieved on moor and acid humus soils through drainage and marling. This, among other things, has resulted in a change of land use. Marling, however, is not a once and for all measure, as the effect only lasts for about 40 years, dependent on drainage and the amount of precipitation. Marling has also been found to have a detrimental effect on plants, and consequently, liming once during the period of rotation is increasingly preferred. Soil productivity is thereby improved.

The areas under grass and green fodder have halved since 1960 (*Fig. 16*), which is a greater proportionate decline than the reduction in the amount

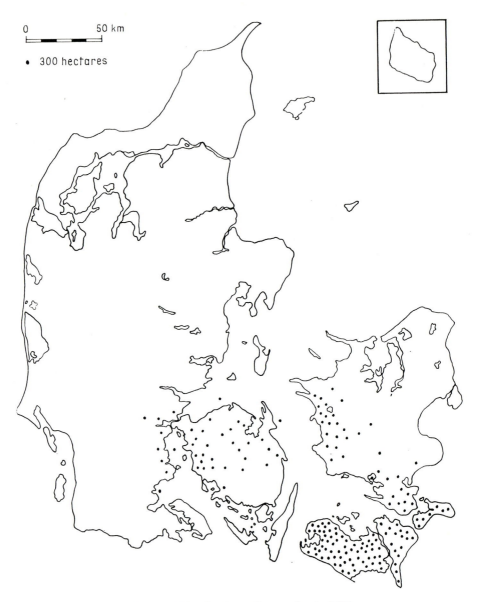

Fig. 15. Distribution of sugar-beet, 1971

Photo 8. Beet field on slightly undulating moraine. In the background cereal fields on a flat-topped hill, built of clayey ice-lake sediments. Fyn (by Kampp)

of livestock; accordingly grass makes up a larger percentage of the feed than earlier. For a given area, grass requires only 1/3 of the amount of fertiliser demanded by a beet crop. Additionally, while it yields only half the calorific value it provides more protein per hectare, and taking costs into consideration yields five times as much per money unit. Grass and green fodder areas as shown in Figure 16 predominate in Jylland, as does grass outside the rotation (*Fig. 17*). The forage harvester saves much work in the grass field.

Figure 18 shows the length of the growing season of some of the more important Danish crops.

The fallow areas are now very few and small in extent.

Crop rotation has been a most effective means of keeping down weeds. But price conditions, rationalisation, and lack of farm labour have forced farmers to adopt a more specialised land use, which has deprived them of the most natural method of weed prevention. The eight-field rotation was most common until about ten years ago (Kampp 1959). Grass and roots were each grown in two fields, while there were four fields of cereals: one winter crop (rye or wheat according to soil conditions), and three spring crops (two of barley and one of oats — in the Islands often three of barley, and in West Jylland mostly one or none at all). But now a 4—5 crop rotation has become common, while even *ad hoc* plans are developed taking the field pattern of the foregoing and the coming years into consideration instead of a fixed crop rotation.

Coastal regions in which low winter temperature and late night frosts

37

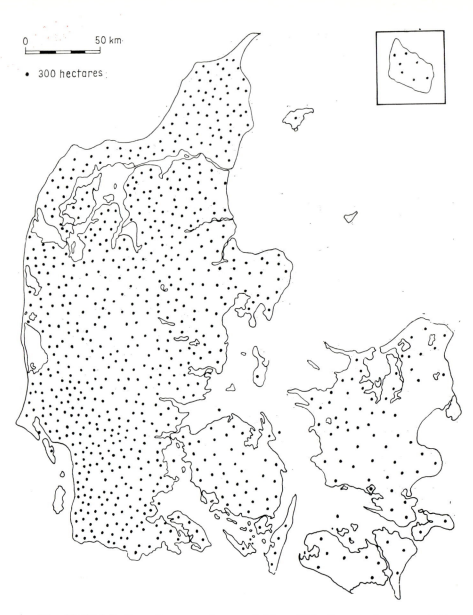

Fig. 16. Distribution of grass and green fodder within the rotation area, 1971

0   50 km

• 300 hectares

Fig. 17. Distribution of grass outside the rotation area, 1971

occur less frequently are suitable for wintering frost sensitive crops and fruit trees. Seeds are grown especially towards the east (*Fig. 19*). They cover only about 3 per cent of the arable land area, but that is more than double the 1954 figure. The most important seeds are the various grasses and oil seeds (rape and mustard), followed by clover and beet seed.

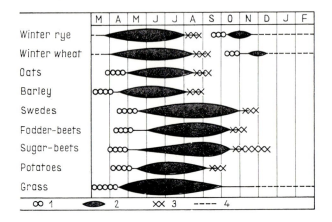

Fig. 18. Length of growing period of crops in Denmark (Kampp 1964b)

1 = sowing; 2 = length of growing period; 3 = harvest, 4 = resting period during winter

Although horticulture is to a certain degree also mechanised and in many places it is being organised by the canneries, it claims more labour than agriculture proper.

The area of horticultural products is 0·3 per cent of the total, and the number of holdings about 10,000, of which 3,400 are under glass. The main area is found in the Islands. 28,000 hectares are used for horticulture of which 11,000 hectares are outdoor vegetables and 12,000 hectares orchards and baccate fruit. Apples account for about half the orchard area.

The production of carrots, tomatoes, and turnip-rooted celery has been increasing slightly. The production of tree-fruit has stagnated since the 1960s. Strawberries are the most important baccate fruit (some 1,600 hectares), and production doubled between 1960 and 1970).

The area under glass is about 600 hectares, and is located mainly near the towns.

The most important limiting factors on fruit growing in Denmark are night frosts during the flowering season, the light conditions, and the temperature in August—September. Accordingly orchards are mainly concen-

40

Fig. 19. Distribution of seed crops, 1971

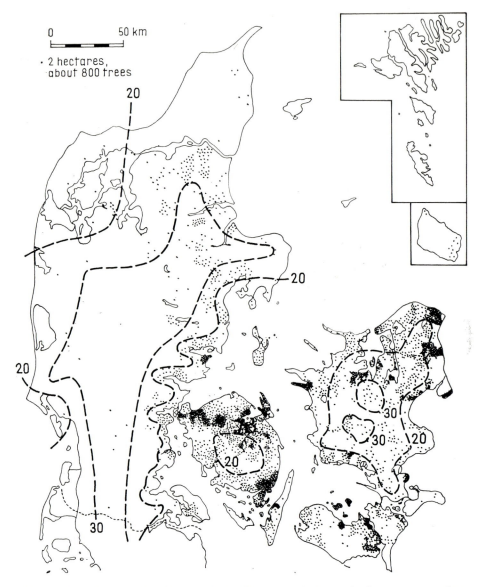

Fig. 20. Distribution of apple trees, 1965. The curves indicate the latest average data for frost after 1 April (Kampp 1969a)

Photo 9. Apple trees in Själland (by Kampp)

trated near the coast where the regional climate is favourable and the soil not too clayey (Kampp 1969a).

Figure 20 shows that the apple orchard areas are mainly concentrated along the coasts of the Islands and East Jylland (*Photo 9*). *Figure 21* demonstrates the distribution of the orchards as a percentage of the total area at different distances from the coast.

Where market gardening and fruit-growing are concerned the proximity of cities is evident, especially where extensive wholesale markets for such produce have been created, i.e. at Köbenhavn, Aarhus, and Odense, although modern transport has diminished the importance of this factor.

The general increase in income and the growing demand from other industrialised countries forced agriculture along new paths. The above mentioned changes in the utilisation of agricultural areas reflect a higher degree of specialisation than is seen in most other European countries. In spite of greater economic risks during years of disease to the barley crop, an increasing number of farms are specialising in a single or at most a few crops and in either a single form of animal husbandry or none at all. The rotation of crops has been replaced by deep tilling, pesticides and herbicides, and special machinery (*Fig. 22*). The establishment of a new trend in the es-

43

tablished occupational and community pattern is, however, hampered by conservatism, and some of the more conservative farmers are still using the old system (*Fig. 23*).

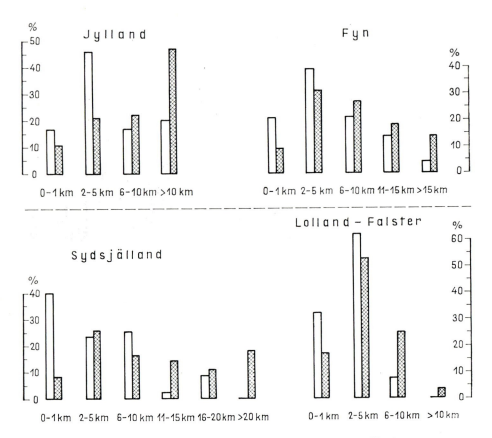

Fig. 21. Distance from the coast. White = percentage of orchards. Hachure = percentage of total area (Dalbro 1967, Kampp 1969a)

CROP YIELDS

Despite the decrease in farm land there was until the 1960s an annual increase in production of 2 per cent because of plant breeding, the extended use of fertilisers and chemicals against weeds and plant diseases, mechanisation, rationalisation, soil improvement, and intimate co-operation between science and practical agriculture, which are closely linked in Denmark.

By international standards the yield per hectare is very high, amounting to an average 4,700 fodder units* per hectare for all crops, which is made up of 4,000 units per hectare for cereals, pulse and straw, 9,800 for roots, and 5,700 for grass and green fodder crops.

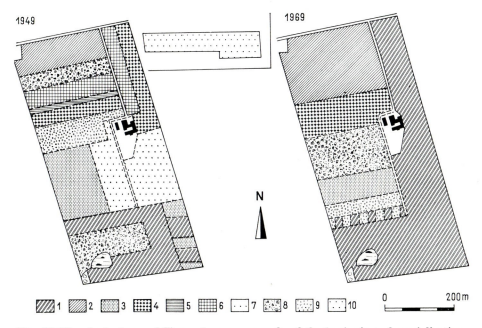

Fig. 22. The single farm of Figure 3 as an example of the beginning of specialisation 1949 = 10 different crops in 14 fields; 1969 = 6 crops in 6 fields (Kampp 1973)

1 = barley; 2 = wheat; 3 = oats; 4 = sugar-beets; 5 = beets; 6 = swedes; 7 = ley; 8 = seeds, 9 = legumes; 10 = permanent grass

Improvements in yield have been highest in the West, due to the improvement of poor soils by means of fertilisers and to a small extent to sprinkler-irrigation. As a consequence, geographical differences are now to a certain extent diminishing.

Yields per hectare have increased by more than 20 per cent since 1950, although total yield for the whole country has been rather stable since 1960

---

* 1 fodder unit = the fodder value of 1 kg of barley, wheat or rye, 1·2 kg of oats, 5 kg of straw, 1 kg of dried potatoes and 1·1 kg of dried fodder roots. Grass and beet-tops are estimated according to their dry matter.

(*Fig. 24*). For cereals and pulses the increase has been rather small, while for roughage a considerable decrease has occurred. As a consequence of the latter the trend in total yield has been slightly downwards since 1964. Silage has remained steady through the years thus compensating for the fluctuations in the production of roughage.

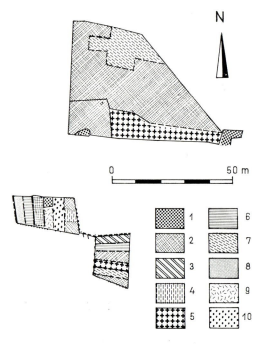

Fig. 23. Three different holdings in the same village. The farm (15 hectares) has only 3 different crops, the neighbouring small-holding (2·5 hectares) only 1, but the outlying small-holding (5 hectares) has 9 crops in 12 fields

1 = buildings and gardens; 2 = barley; 3 = wheat; 4 = oats; 5 = sugar-beets; 6 = beet roots; 7 = rotation grass; 8 = seeds; 9 = legumes; 10 = permanent grass

From 1960 to 1973 the share of barley in grain production grew from 58 to 82 per cent, and barley now accounts for some 45 per cent of total feed units; the remainder is made up of grass, root crops, and straw. In 1973 the country became totally self-sufficient in grain (Thomsen 1974). Root production has fallen to half of what it was in 1960; 90 per cent of total crop production is used for feeding stock. The total production of fruit and vegetables was not subject to any decisive change during the 1960s.

Agricultural development is of course not restricted to farming activities but is closely associated with industry, particularly for the supply of agricultural inputs and for the processing of agricultural products.

The future will show if Denmark's membership in the EEC will lead to any change in land use and production.

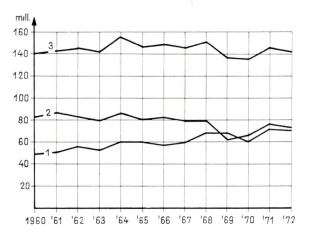

Fig. 24. Harvest yield in millions of crop units 1960—1972 (Thomsen 1973)
1 = cereal and pulse; 2 = roughage; 3 = total

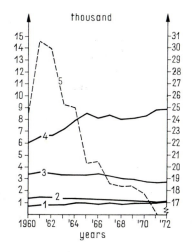

Fig. 25. Livestock 1960—1972 (Thomsen 1972)

Left scale: 1 = sows and boars; 2 = dairy cows; 3 = cattle, total; 4 = pigs, total.
Right scale: 5 = poultry

47

ANIMAL HUSBANDRY

Livestock is a very sensitive branch of production and is intimately linked with economic conditions (*Fig. 25*).

The main breeds of cattle are Red Danish Dairy Cows (*RDM*) and Black and White Danish Dairy Cows (*SDM*), but Jersey Cows are becoming increasingly common (*Photos 10, 11*).

During the last 30 years the number of horned cattle has continually decreased. At the beginning of this century the distribution of dairy cattle was oriented mainly towards the east but its centre of gravity is now in the west (*Fig. 26*), because the farmers of the Islands and East Jylland have to a large extent given up dairy farming, with the result that Köbenhavn must increasingly buy liquid milk from Fyn and Jylland.

The number of holdings without polled cattle now amounts to more than 30 per cent of the total (Thomsen 1974).

As each dairy cow must be supported by 1/2 hectare of crop land, the reduction in the number of herds thus frees land for cereal growing.

During the same period the structure of the herds has changed. The average number of dairy cows in the herds has decreased from 50 per cent in 1939 to 40 per cent today, herds with 15—30 dairy cows now dominating.

Although the total number of dairy cows has fallen by 25 per cent since 1960, milk production has dropped by only 15 per cent, as the yield per cow has increased to 4,046 kg (*Table II*).

South Jylland comprises the narrow part of Jylland from the German border to a line running approximately from the narrowest part of Lille Belt in the east to a point a little south of Esbjerg in the west (see Fig. 1).

In Denmark most cattle herds are kept for dairying with beef as a by-product. At present 6,000 of the 135,000 holdings concentrate exclusively

Table II

	1909	1939	1951	1962	1971	1939—71
Cattle, total	2,254,000	3,326,000	3,110,000	3,504,000	2,723,000	
Dairy cows	1,282,000	1,642,000	1,584,000	1,408,000	1,105,000	
Total milk production, million kgs	3,462	5,277	5,233	5,355	4,556	—14 pc
Average milk production kgs/per cow	2,700	3,213	3,304	3,802	4,046	+26 pc

Photo 10. Red Danish dairy cows with horns removed on rotation grass behind an electric fence. Mön (by Kampp)

Photo 11. Black and white Danish dairy cows in a littoral meadow. Jylland (by Kampp)

Fig. 26. Distribution of dairy cows, 1971

Photo 12. Beef cattle, Hereford breed, grazing all year round on permanent pasture. Klintholm Manor, Mön (by Kampp)

on beef (*Photo 12*). As a consequence of the close connection between the production of beef and milk it is difficult to sustain the former when the latter goes into decline.

The number of calves for veal has increased from 26 per cent to 34 of the total stock (*Table III*) because they are now slaughtered at a later period of life.

Table III

	1909	1935—39	1950—54	1960—64	1971	1935—71
Percentage of dairy cows in the stock	56·9	51·2	49·4	41·9	40·5	
Percentage of calves	21·6	25·5	26·3	32·7	34·9	
Dairy cows in East Denmark	823,000	862,000	784,000	659,000	432,000	—49 pc
Dairy cows in NW and S. Jylland	459,000*	747,000	742,000	775,000	672,000	—10 pc

* Minus South Jylland, which then belonged to Germany. The number in 1920 was 91,000.

4*

Fig. 27. Distribution of pigs, 1971

Fig. 28. Distribution of poultry, 1971

Bullocks are as formerly kept by farmers in South and West Jylland.

From most farms the milk is taken to the nearest co-operative dairy. The first Danish co-operative dairy was founded in 1882, but the number grew quickly. The total reached its highest point in 1935 with 1,404, since when the number has decreased because of rationalisation through the amalgamation of dairies. At present there are about 260 co-operative dairies in Denmark.

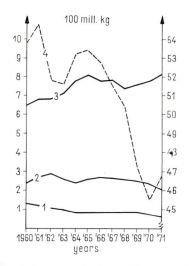

Fig. 29. Animal production (Thomsen 1972)
Left scale: 1 = eggs; 2 = beef and veal; 3 = pig meat and pork. Right scale: 4 = milk

In 1952 the Danish Ministry of Agriculture ordered that dairies using the trade mark for first class Danish agricultural produce — the Lur* — on butter and cheese should accept milk only from herds which had been registered free from tuberculosis. It was the dairy industry, i.e. the farmers themselves, who asked the Ministry to issue the order and the industry itself keeps the register. Bovin tuberculosis has now been eradicated throughout Denmark to the good of the public, the stock, and the producers. The measures instituted to control the herds are so effective that there is no reason to fear any recurrence of tuberculosis among Danish cattle.

Pigs are fairly evenly distributed throughout the country with a small predominance towards the east and in North-West Jylland (*Fig. 27*). The

* The old Danish Bronze Age trumpet.

54

keeping of pigs is easier to rationalise than that of dairy cows, and the number of large pig enterprises is growing; enterprises with more than 100 animals now account for more than 26 per cent of the total (Thomsen 1974).

The co-operatives have a 92 per cent share of the market. Half of the pig production is turned into bacon and exported mainly to the United Kingdom, which is naturally done co-operatively. By-products from the bacon factories, such as offal and pork, are sold either direct by the primary co-operative factories or by joint factories. Quite a lot of canning is done jointly.

Sheep are not a very important branch of Danish agriculture; they are kept mainly on the marshy area of South Jylland, and tending them claims relatively little effort.

Poultry is now kept on only 45 per cent of the holdings and is mostly concentrated near the egg packing stations and poultry-slaughterhouses (*Fig. 28*).

The output from animal husbandry, especially pork and bacon (*Fig. 29*) has increased more rapidly than crop production because of a growing import of oil cake, the above mentioned increase in crop production, and the continuous improvement of stock and feeding practices.

D. AGRICULTURAL REGIONS

In some countries regional geographical differences are immediately apparent. It may be a question of different elevation, for instance low lying plains, table land and mountains. There may be different soil conditions: rocky ground, sand, gravel, clay, or peat. Or there may be differences of temperature in countries such as Chile and Norway which cover vast distances in a north—south direction. Again, it may be a question of variations in precipitation creating deserts in one region of the country and densely wooded areas elsewhere. Variations in population can account for differences as well.

Even small countries may exhibit considerable disparities. Luxembourg covers an area only one seventh that of Denmark, but the northernmost third of the country — the dissected, well-timbered Ardennes plateau — houses only one tenth of the population while nine tenths live in the southernmost region — the undulating lowlands of "Le bon pays" — comprising two thirds of the entire territory. The slate-mountains of the Ardennes form an impermeable shield preventing the formation of groundwater, and the loose top soil is mostly thinly spread and not particularly fertile. Different altitudinal zones result in considerable differences in temperature and precipitation.

Compared with most other countries Denmark does not exhibit regional differences of any consequence in respect of soil, terrain, or climate. And yet in a single year the yield per hectare in a given parish may range from 6 to 60 hecto-kilogrammes for wheat, 8—42 for rye, 8—48 for oats, 35—500 for potatoes, and 50—1,000 for swedes, the extremes representing crop failures and bumper harvests respectively. The variations are due to several phenomena such as differences in soil conditions, small climatic variations, and simply the general caprices of the weather.

Thus the geographical conditions of soil and climate, the winter hardiness of the various varieties, together with prices of farm produce, decide the choice of crops within the system of rotation.

Soil quality is a function of the physical, chemical, and biological state of the soil. Only a few of the factors decisive for soil fertility lend themselves to numerical presentation — let alone to representation on maps.

In the first place, soil analyses undertaken in various parts of the country are not carried out systematically and, secondly, they do not consider, by any means, all relevant soil factors. The qualitative assessment formerly used as a basis for taxing Danish agricultural areas, for instance, was merely an estimate of the physical conditions and disregarded the chemical and biological factors. These in turn depend on climate and the various methods of cultivation which, of course, are of vital importance. But until a map of landscape ecology has been prepared decisions regarding the most profitable land use can only be a matter of practical experience.

During the years following 1939 a research project was started with the intention of subdividing Denmark into agricultural regions according to quantitative principles (Kampp 1944).

Despite the predominance of animal husbandry in Danish farming practice the distribution of stock, converted into livestock units per unit area, could not form part of the basis of the classification because it would have created a bias discriminating against parishes emphasising seed-growing, and sugar-beet. Besides, livestock would have been unduly dominant over soil quality in, for example, the river valleys of Jylland. Poultry-farming, too, would have created a minor bias, since chicken fodder is seldom grown locally, and other types of animal husbandry in the poor soil districts of Jylland are based mainly on imported feeding-stuffs and to that extent independent of soil quality. In other parts of the country some livestock fodder is grown locally and this would have counted twice if the stock also had been taken into consideration in the final regional division.

Consequently the delimitation of agricultural regions had to be determined on the basis of soil quality alone, although a detailed map of soil quality covering the whole of Denmark did not exist.

Since the conditions of cultivation inherent in soil quality are so numerous and partly unknown, and since, as mentioned above, only a few of them lend themselves to numerical representation, it was necessary to find an indirect measurement of the influence of soil factors on quality, i.e. an assessment of productivity or yielding capacity. This method was not unlike that of the hartkorn* assessment for the 1844-register which was made for reasons of taxation and was an estimate of the net yield of well-tilled

---

* Hartkorn is a Danish standard of land valuation. It was based on the normal yield of the soil, and 3,575 hectares of the most fertile soil constitute 1 tönde hartkorn. On an average, however, 1 tönde hartkorn required 5,995 hectares in the Islands, 14,630 hectares in Jylland, and 9,900 hectares on average throughout the entire country. Up to 1903 taxation was based on the amount of tönder hartkorn.

soil, which eliminated the extra profit that might show up on land value maps of farm land in close proximity to towns.

As the most careful chemical and physical soil analysis would in all probability never be an exact indicator of soil quality, the crops themselves were used as a surrogate for this.

As a preliminary step two maps were drawn up (Kampp 1944b). The first represented the geographical distribution of average yields of the most important Danish crops over a three-year period, which indicated to a certain extent the yielding capacity of the soil (*isodone* map*). But as yield is not exclusively the result of soil quality, it was considered necessary to add a further factor as a control in the process of sub-division.

Because wheat and at that time also barley required high quality soil, these crops were judged to be a suitable basis for classification and a second map was drawn up (*isodense** map*).

The area of wheat and barley as a percentage of the total area under rotation in each parish was found to be markedly different in various parts of the country, but at the same time fairly homogeneous over large districts. Isorithmic maps based on the wheat and barley distribution for the years 1838, 1907, and 1939 turned out to be largely similar. The sub-division of the country based on this material almost corresponded with that suggested by the isodone maps and the division of the country into seven agricultural regions was therefore carried out with some confidence (*Fig. 30*). Isodone and isodense maps for 1946 and 1962 did not differ essentially from the original ones, which proves the stability of the regionalisation.

As mentioned above the distribution of animal husbandry was not considered in devising the regional sub-division, but investigations for the period 1907—46 nevertheless showed a large degree of correspondence between the number of pigs and dairy cattle per area unit and the sub-division; also the yield of milk per cow was lowest in region I, and highest in region VII (Kampp 1959a).

REGIONAL GEOGRAPHICAL DISPARITIES

In this description of the seven regions soil quality is lowest in region I and highest in region VII.

---

* iso- (Greek) = equal, -done, from Latin: dono = I give.
** -dense, from Latin: densitas = density.

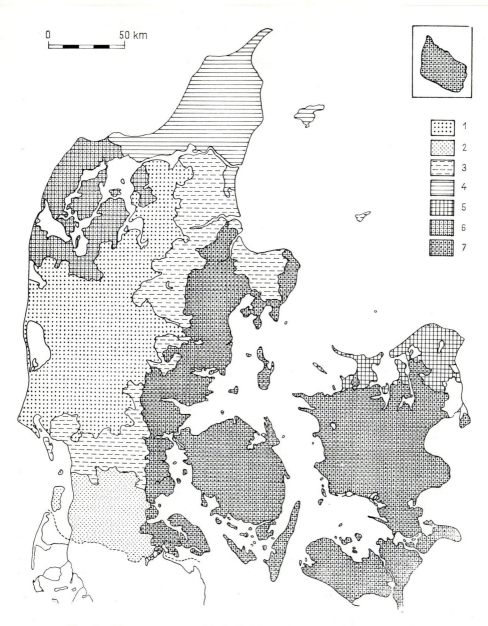

Fig. 30. The agro-geographical division of Denmark (Kampp 1959b)

1 = West Jylland; 2 = West South Jylland; 3 = transitional zone; 4 = Vendsyssel; 5 = North Själland; 6 = the West Limfjord region; 7 = East Denmark

I. THE AGRICULTURAL REGION OF WEST JYLLAND

Morphologically this region is characterised by smooth undulating hill-islands and melt-water deposits with broad stream-valleys and deltas.

For centuries the greater part of the surface was covered with heather. The moor was a by-product of past human activities — a result of thoughtless forest clearing, slash and burn methods of cultivation, which exhausted the podsolised soil and prevented the rejuvenation of the natural vegetation.

From about 1860 much land reclamation was undertaken, above all the draining of bogs and meadows and the cultivation of the moors. Today six sevenths of the moorland has been converted into agricultural land.

Spring sown cereals and beets have to a large extent replaced winter crops and grass, thus leaving the fields bare during spring gales with the result that wind erosion has set in across areas of sandy soil. The areas showing the greatest increase of farmland seem to co-vary with the areas hit most badly by wind erosion. These are the same areas that were abandoned during the Iron Age when people knew no better than to vacate the plot if soil erosion set in. Today there are better means available thanks to the possibilities of rational planning, e.g. windbreaks in the main running north-south as a protection against soil erosion caused by the prevailing westerly wind. The dominant trees are conifers both in the open country and in the many large plantations which frequently have replaced farmland of doubtful agricultural value.

Taking into consideration forestation and the retreat from previously cultivated outposts, the expansion of the cultivated area is still much larger here than in any other region, although the percentage of farmland is smaller than elsewhere apart from North Själland.

Formerly rye was the dominant cereal, but today by far the largest part of the area is sown with barley which has taken the lead in this region as elsewhere in the country. Oats and mixed grain are grown in relatively small areas while there are even smaller areas of rye and next to no wheat.

Potatoes are grown here for fodder and therefore occupy a larger percentage of the area than in other regions. Swedes are the dominant root-crop while seed-growing is almost nonexistent.

In spite of a larger increase in numbers, population density in the region is still much lower than elsewhere in the country.

In earlier times the farms were unsheltered from the wind and were mostly situated far apart, probably because holdings of considerable size were necessary to support a family. Indeed holdings still exceed the average size for the country as a whole. Black and White Danish Cattle (*SDM*) are the dominant breed.

Until recently an 8-field rotation was the rule, although 7-field system existed, where livestock was numerous and a 6-field where many potatoes were grown. The use of the fields, but hardly their number is determined by grass outside the system of rotation. These natural grass-areas are, however, receding now that the meandering streams have been straightened out into canals and the wide brims have been drained as preparation for conversion into cultivated fields.

During the first half of the present century the number of small-holdings in West Jylland increased from 15,000 to 35,000, but large stretches are still completely devoid of manors and villages. In a few places rather large village communities are found, associated with stream-valleys, secondary roads and railway stations.

The cultivation of the Jylland moors is striking evidence of man's capacity to manipulate his environment to suit his needs. The frugal life of the moorland peasants while laboriously cultivating the heath is legendary.

II. WEST AND MID-SOUTH JYLLAND

Region II comprises sandy moorlands and rather clayey hill-islands plus marshes and many meadows.

In few places in Denmark has physical geography played such an important role in the choice of sites for settlement as in the marshes. The erection of a farm on artificially raised ground named a "wharf" (from which the word "*värftsgård*", "wharf" is derived) dates back to the time before dikes were built as a bulwarks against floods. Such farms were mainly isolated habitations.

Later on buildings were placed along the inner banks of the outer dikes as well as on both sides of the inner dikes. In modern times farmhouses can also be seen on the marshland proper.

Besides the "wharves" and the dikes, other man-made features characteristic of the marshes are seen, such as drainage canals and hollows where the material for "wharves" and dikes has been dug out.

Villages are few and far between in the marshes whereas they are more frequent on the rim of the geest-land.*

* Geest is a generic term for the older elements of the landscape adjacent to the marshes. From Old Frisian gast = barren, dry, highlying.

In the remaining part of the region habitation is scattered and scarce, although typical villages and road agglomerations exist.

Compared with the rest of Denmark there was a large number of small-holdings and large estates in South Jylland during the time of German rule, but relatively few medium-sized holdings.

Parcelling-out did not begin until after 1920 when South Jylland was returned to Denmark after World War I and consequently the size of State small-holdings is larger here than in the rest of the country. Also the average size of other holdings is larger in this district.

When South Jylland again became Danish territory not only were the glebes and the areas, which during the years of German sovereignty (1864 to 1920) had been taken over by the State, parcelled out, but in the following years the Central Land Board bought up a great many scattered lots with the objective of consolidating adjoining areas for sub-division.

When this goal had been attained the preparatory work of improving the soil by drainage, repairing and enlarging the road system, erecting farm buildings and houses for people with secondary and tertiary trades, building schools etc. was organised so that whole groups of people could move into "ready made" villages.

During the years of Prussian rule the swampy areas had been left to themselves, but now drainage on a large scale was carried out and as a consequence there has been a major extension of arable land in most parts of the region.

The hectare-yields are rather small. The percentage of grain area is lower here than in all other regions; oat after barley is the dominant crop in this region of high rainfall.

Grass areas play a more significant role within the system of rotation than elsewhere in the country. The same is true of grass outside rotation, which is substantially increased especially in the flat marsh areas in the west, drained by straight, parallel ditches. The extensively cultivated marsh where settlements are placed on the geest, is still used mainly for grass-growing, whereas cereals and roots are strongly on the increase in the more intensively cultivated areas of the marsh.

A flexible system of rotation is the rule because conditions of soil quality and humidity vary widely. Many holdings use 2 or 3 different rotations, e.g. an 8-field rotation on morainic soils and one or two 6-field rotations on cultivated marsh soils.

In many places within the district an apparently inappropriate distribution of scattered plots is seen, a remnant of the open-field system. Enclosure was here begun earlier and therefore not thoroughly carried out.

This, however, is not of any serious consequence because a considerable part of the marshy area is used for the fattening of bullocks.

Wooded areas are very small, mostly coniferous; towards the east there are a few deciduous groves.

III. THE TRANSITIONAL ZONE

This zone is transitional between regions I and VII, in respect to both density of population, size of farming areas, quality of soil, and distribution of practically every cultivated plant, breed of cattle, and type of forest. Beet-roots prevail in some parishes, and swedes in considerably more.

Villages are prevalent and fairly evenly distributed.

It is one of the regions of Denmark where the parcelling-out of farmland in the first half of this century has left very conspicuous marks on the landscape. Several groups of neighbouring small-holdings uniform in size have been parcelled-out, partly on private initiative, partly through small-holders' associations and partly by the Central Land Board.

IV. VENDSYSSEL

In Vendsyssel the interaction between man and the natural environment is more marked than in most other parts of Denmark. The village districts of the western Yoldia-plains and rather even morainic areas are clearly distinguished from the very scattered habitations of the raised Litorina sea-bed, the Yoldia-plains and morainic hills towards the east. The villages of Vendsyssel are mostly small compared with those of Eastern Denmark.

From the coastal dunes in the west, sand has shifted in the course of time across the old cultural features, especially towards the south-west, where the Litorina-expanse was the largest, and on the spit of Skagens Odde which is completely covered with blown sand.

In this extremely windy climate (see Photo 1) windmills were erected at practically every farm during the years from 1910 to 1930. They are all gone now, being replaced by electricity.

Root crops in order of significance are as follows: swedes, potatoes and beet-root. Practically no fruit or seed is grown.

A 7-field rotation using three or four fields for grain, one to three for grass, one for beet or swedes and one for potatoes is common.

SDM is the dominating breed of cattle.

Store Vildmose was the largest high bog in Denmark. It was situated on a large Litorina-plain in the south-western part of Vendsyssel. Until the beginning of this century it was a nearly impenetrable swamp, dotted with small ponds and bushy with large continuous plains of peat, covering Iron Age fields and settlements.

In former days villages bordering Store Vildmose were mostly situated along the edge of the morainic land surrounding the high bog, which was thus well situated for grazing cattle being turned out there in the morning and taken home again at night. But in 1921 agricultural improvement was started. Roads were built, drainage canals dug, and the bog surface accordingly sank by about two meters. Subsequently large areas were deeply ploughed and after marling sown with grass or cereals. In consequence as early as the 1930s the greater part has been converted into agricultural land. On the nearly circular plain of about 7,000 hectares many thousands of cattle now graze and cultivated fields leave their mark on the landscape round the Central Farm and the many homefarms and rented holdings. Only a small area is to a certain degree still bog.

V. NORTHERN SJÄLLAND

Northern Själland is a very densely populated region, and is increasingly becoming a commuter-region as well as being an area of summer residences for the inhabitants of Köbenhavn. In the past isolated holdings were a characteristic feature, but during the present century there has been a marked decline in the area farmed so that today little more than half the area is agricultural land which is a far smaller percentage than in other regions. Particularly grass outside the rotation occupies a minimum area.

Red Danish Dairy Cattle (*RDM*) is the dominant breed.

On the predominantly sandy loam a 7- or 8-field rotation is usual. The percentage of seed areas is only slightly smaller than in region VII. Hectare-yields are average.

Commercial fruit growing is far more important in this region than in any other. Fruit growing dominates the scenery partly because this region has the smallest average size of holding, and partly because of the extensive stretches of light soil. Many people choose fruit growing as a neat and clean side-line in addition to urban employment (Kampp 1969a). Gardening and fruit growing also benefit from the vicinity of the Capital on the east coast. As south-east winds are not very frequent wind-borne air pollution affecting the fruit and garden areas is not a serious threat. Additionally, in the coastal regions there is less risk of frost damage during the flowering season, while the proximity of the sea generates favourable temperatures during autumn when the fruit is ripening.

In terms of area, region V has more woodland than other parts of the country, because among other things considerable parts of the region were once Crown land and the hunting domaine of the Kings of Denmark. The forests of North Själland are gaining in importance today as recreation areas. Beech is the most common species, but conifers occupy one quarter of the wooded areas, being particularly dominant in the dune plantations in the northernmost part of the region.

VI. NORTH-WEST JYLLAND

The western part of the Limfjord-region is bordered on the northwest by dunes, now covering formerly cultivated lands on fertile, predominantly clayey, young morainic soils. As regards soil this region is almost as favoured as region VII but springs and summers are colder and more windy. Barley and oats are the most important crops of the region, while swedes are the dominant root crop. Grass, too, plays an important role in the rotation. A 6-field rotation is common consisting mostly of three fields of cereals, two of grass and one of roots.

In the dune areas to the northwest the number of deserted farmhouses is steadily growing.

Dairy cows and bullocks are predominantly *SDM*. Plantations are mostly coniferous.

Habitation is often road-oriented, and rather large villages are as a rule clearly separated from the scattered farms.

VII. EASTERN DENMARK

Farm land, in spite of a decrease during the present century, occupies about 80 per cent of the area of this region. The soil is on average of better quality than that of any other region. While barley is the main cultivated cereal, the dominant root is beet-root. Seed-growing covers a larger area than in any other region. Most gardens have fruit trees, mainly apple trees, and there are numerous commercial orchards. Grass takes up far smaller areas than in any other region.

Most of the dairy cows are *RDM* or *Jersey*.

An 8-field rotation has been the rule, but the rotation of individual farms varies, for instance, with the number of livestock. On holdings without cattle, grass and root crops are of no importance except sugar-beet for the factory. There is a tendency to enlarge the fields as much as possible to facilitate the better use of machinery (see Fig. 22). Where permanent grass is abundant it may remain outside the rotational cycle. Often the rotation provides merely a loose framework within which variations are possible, especially as far as industrial crops and seed crops are concerned, according to soil quality and market conditions.

Settlement is dense especially along the roads and in many places one village can hardly be distinguished from the next. The single, isolated holding is characteristic of parts of Mid Själland and West Fyn.

In Bornholm there are neither manors nor villages. It forms an island of small-holders, but as a consequence of, among other things, marked emigration amalgamation of holdings takes place here more frequently than in the rest of the country.

Precipitation is lower than in the greater part of Denmark because many depressions tend to collapse before reaching the area. Spring especially is unusually dry, but the cold, early summer brings high relative humidity and accordingly low evaporation.

Excellent road and railway systems exist in the region. Because of man's intervention woodland is mainly of scattered beech or beech-plantation. The oak was more valuable for building houses and ships, while the hogs preferred acorns as food — all of which served to decimate oak-woodland. In addition, beech-woodland is nowadays preferred for recreation.

The stability of the division

It has been demonstrated (Kampp 1970c) that the regional classification described above has exhibited a high degree of stability through time.

Investigations made at intervals of a few years may, of course, show different results. Thus various crops and animals were, as mentioned, differently distributed in the past, a result of efforts to select crops giving the highest financial yield per unit area. This means, for example, that the economic results of experiments with different strains of cereals can change the entire pattern of crop distribution.

Modern Danish agriculture is becoming increasingly independent of the natural environment and any prognosis must anticipate a continuous improvement in the quality of the soil in the poorest agricultural regions. Apart from extremely hard winters which may destroy the winter wheat crop, the weather will not generally influence per hectare yields. Not even changes in groundwater-level in the meltwater-valleys caused by variation of rainfall have much influence on the utilisation of grain areas. Conceivably, new methods of cultivation could alter the land use pattern, for instance, the more intensive use of fertilisers. Alternatively the soil might be improved to the level prevailing in a neighbouring region. The drainage, irrigation, and manuring of former tunnel- and meltwater-valley floors have improved their state of cultivation. Also part of the sandy morainic soils has been improved. But none of these changes has altered the geographical pattern of isodones and isodenses, the fundamental basis of the division.

The initial statistics on which the division was based were data for the years 1939 and 1946 for which specified statistics were available from each of the 1,800 rural parishes of Denmark. Today statistics are only calculated for the 275 municipalities into which the country has been divided since 1970. Consequently it is no longer possible to draw maps as detailed as those on which the division was based.

Nevertheless, *Figure 31*, based on data for 1971, permits the conclusion that the geographical isodense pattern has changed very little even though there has been a strong upsurge in wheat and barley as a percentage of the total rotation area. This progress is to be seen from the signature columns (*Table IV*) which are determined by the distribution curves, and show much larger figures for 1971 than those found for previous years.

Table IV

(in per cent)

	1907	1939	1962	1971
1.	$\leq 1\cdot 4$	≤ 9	≤ 24	≤ 44
2.	$1\cdot 5 - 5\cdot 9$	$10-19$	$25-35$	$45-50$
3.	$6\cdot 0-11\cdot 9$	$20-27$	$36-47$	$51-56$
4.	$12\cdot 0-16\cdot 4$	$28-33$	$43-47$	$57-60$
5.	$\geq 16\cdot 5$	≥ 34	≥ 50	≥ 61

For 1—5 in the left column see *Fig. 31*.

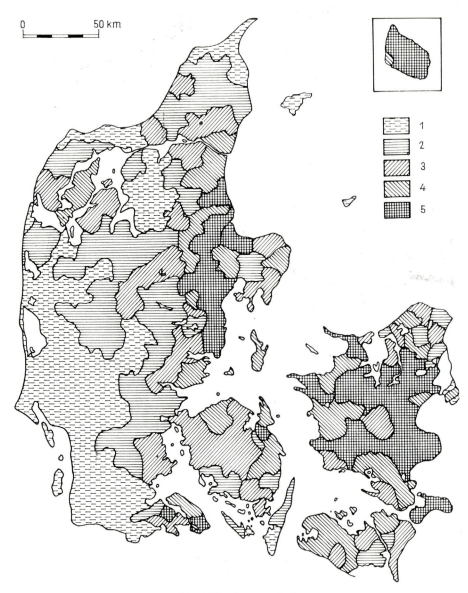

Fig. 31. Isodenses, 1971

1 = ≦44; 2 = 45—50; 3 = 51—56; 4 = 57—60; 5 = ≧61

A simplified division

If the character of agriculture is to be depicted on a small-scale map for example, for Europe or even part of Northern Europe, the scale used will be such that it is not possible to represent the seven agricultural regions into which Denmark is divided. Under such circumstances a simplified division is both necessary and sufficient to portray the agricultural differences within the country.

Table V

Districts	Pigs per square kilometre		Dairy cows per square kilometre	
	agricultural area	rotation area	agricultural area	rotation area
W	253	314	42	54
M	264	324	40	48
E	290	334	30	34

For such purposes a simplified sub-division into three districts is proposed: regions I and II are amalgamated to form the West Danish District named *West (W)*, regions III and IV to form the Middle District named *Mid (M)*, and regions V, VI, and VII to form, the East-Danish District termed *East (E)*. Whether the regionalisation is based on seven agricultural regions or on three major districts it reflects the basis of agricultural production over many centuries.

Even for Denmark alone the threefold division can be useful if only a general survey is required as in the example shown in *Table V*, presenting a *W — M — E* division of the number of pigs and dairy cows in 1971.

E. ECONOMIC INPUTS AND OUTPUTS

At the beginning of this century some 40 per cent of the total population was occupied in agriculture, which today has dropped to about 9 per cent. However, during the same period, the number of man-hours to work one hectare of wheat has fallen from 120 to 20, while the annual output per man is now 2—3 times higher than 15 years ago.

The early 1930s saw the highest number of persons occupied in agriculture, when the figure was 500,000 men per year.

The rapid decrease to the present has, among other things, been due to amalgamations as mentioned in chapter C, but particularly to the flight of agricultural labour to the towns (*Table VI*). As seen from *Figure 32* the number of hired labourers is smallest in East Denmark, where industry has acted as a magnet, drawing hired labour to urban districts.

AGRICULTURAL LABOURERS

From Table I it can be deduced that the number of farm owners has fallen by about 30 per cent since 1960. Together with the reduction of the number of hired helpers during the same period it can be seen that the total number of people occupied in agriculture has been almost halved since 1960.

MACHINERY

A consequence of the revolutionary reduction in the labour force has been an immense rise in the number of farms where the farmer must do the daily work alone, perhaps with some casual hired help. Out of the present number of 135,000 holdings, at least 120,000 are now family-farms. During the same period they have also increased in average size, which has only been made possible by a corresponding rise in the use of machinery (*Table VII*). There has been a growing demand for larger and more powerful tractors and combines, even though such heavy machinery has the drawback espe-

Table VI

	1960	1971	1972	1971 as a per cent of 1960	1972 as a per cent of 1960
Men:					
Children and relatives	28,510	8,726	7,535	31	26
Regular hired workers, unmarried	50,054	9,187 ⎫		18 ⎫	
Regular hired workers, married	14,448	7,425 ⎭ 16,269		51 ⎭	25
Total	93,012	25,348	23,804	27	26
Day-labourers, casual etc.	5,530	3,794	4,927	69	89
Women:					
Children and relatives	11,656	1,685	1,409	14	12
Regular hired workers, unmarried	14,936	1,151 ⎫		8 ⎫	
Regular hired workers, married	2,273	687 ⎭ 1,626		20 ⎭	9
Total	28,892	3,523	3,035	12	11
Day labourers, casual	885	1,877	1,500	215	169
Grand total	128,319	34,542	33,266	27	26

Table VII

Agricultural machinery

	1960	1965	1970	1972
Tractors	111,300		174,639	172,844
of which diesel tractors			101,411	106,562
Combines	8,900		42,253	43,669
Forage harvesters		38,583	53,415	

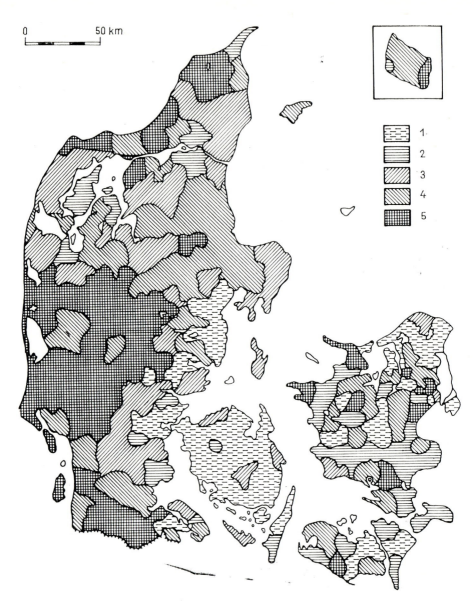

Fig. 32. Farm hands per 100 hectares of agricultural area, 1971
1 = ≦0·8; 2 = 0·9; 3 = 1·0—1·1; 4 = 1·2—1·4; 5 = ≧1·5

cially in East Denmark, of compacting the soil and impairing its moisture-retaining properties. The number of tractors and combines rose by some 63,000 and 33,000 respectively between 1960 and 1970. As saturation point is approached the rate of increase has been slowing down.

The use of machinery has had the consequence that very few farmers have work horses any longer. Thus 250,000 hectares of land formerly used for horse forage are now available for producing pig feed, vegetables and other crops.

FERTILISERS AND MANURE

As a consequence of the growing cereal area and the decreasing amount of natural manure, fertilisers have become increasingly important (*Table VIII*).

Table VIII

Total use of fertilisers	1961/62	1966/67	1970/71	1971/72
N	133,600 tons	215,200 tons	289,300 tons	308,300 tons
Ph	49,700	56,500	55,300	58,100
K	142,100	154,900	150,900	158,400
Total use of manure				
N		153,500	138,300	139,300
Ph		52,900	48,400	48,800
K		174,600	146,600	144,400

TRENDS IN AGRICULTURAL PRICES

The price development

Compared with conditions during the first part of the 1950s there was a total increase of 37 per cent in the prices for agricultural products as a whole, although the increase in many input prices was much larger (*Table IX*).

Table IX

1950/51—1954/55 = 100

	1960/61	1971/72
Livestock products	98	142
Vegetable products	90	110
Agricultural products	96	137
Oilcake	92	126
Fertilisers	105	119
Building works	120	274
Personal properties	128	188
Farm hands' wages	140	487

TRENDS IN ECONOMIC OUTPUT

Table X

Value of production and factor value by farmer (Stat. undersøgelser 1969, in mill. Kr.)

	1960/61	1971/72
Sales of vegetable products	781	1219
Sales of animal products	6075	9582
Changes in livestock and stocks	+262	+37
Total	7118	10,838
Expenditure on raw materials and auxiliary materials	2309	3504
Gross income	4809	7334
Gross income + general support	4809	7613

Finally, we present the total output of Danish agriculture by crops and animal products (*Table XI*):

Table XI

The total output of Danish agriculture, 1971
(in thousands of metric tons)

Quantity of crops:

grain, total	7,026
winter wheat	447
spring wheat	138
winter rye	139
spring rye	11
barley	5,458
oats	701
mixed grain	132
straw, not burned	5,294
pulse	80
roots	13,115
potatoes	750
sugar-beet for factory	1,999
sugar-beet for fodder	1,004 ⎫
fodder sugar-beet	6,665 ⎬ 7,669
mangels etc	281
swedes	2,416

Production of the more important livestock products:

milk	4,557
butter	124
cheese	120
eggs	75
beef and veal	231
pig-meat and pork	816
poultry meat	80
fat	55

F. THE FUTURE

On the preceding pages an account has been given of the present structure of Danish agriculture. A prognosis concerning the years to come is more difficult to construct because more than ever before development is so rapid that what was thought to be a safe prediction a year ago may have been replaced today by a completely different set of problems. National planning, regional development, market conditions, prices — they all change. And what is to be done in an energy crisis, when there are virtually no horses left? As to prices Danish farmers rarely leave a particular enterprise on account of temporarily falling prices. In industry production can be shelved for a while and then resumed when stocks are exhausted. In farming, horticulture and fruit-growing, there are only limited possibilities of doing this, and it is of course not possible to stop producing milk for a month or so.

Intensification of farming on the whole has reached a peak, and as has been said above, the tendency is about to be reversed. There is general agreement that the future goal must be increased production per person and not per hectare. The development of the small holdings is not propitious, at least not for the present. The present structural change has first and foremost been a question of amalgamating holdings. Up till now joint operation is more than twice as widespread in the east as in the west.

This increases the possibilities of coordinating production through machine-pools and joint enterprises. In this way rationalisation until recently restricted to the industries processing agricultural produce is already in some places spreading to the very heart of farming, part of the work having shifted to industry, e.g. ready-made fodder and fertiliser-compounds.

There is a growing realisation that in agriculture specialisation will provide the basis for cheap production as in industry. Experience from industry teaches us that efficiency and the rentability of an enterprise hinge on distributing overheads over the largest possible output.

Attempts at specialisation have already been made with joint technical assistance, joint experimental stations and exploitation of waste-products. Potatoes for consumption grown on the island of Samsö, and carrots from the Lammefjord district are washed, sorted and transported jointly. Cultivation of green peas, previously a small-holder's crop, is becoming a large-scale undertaking. The canned-goods industry is concentrating pea-growing regionally, because the heavy machinery used in the industry is specially designed for large fields. This gives the factor the possibility of managing the entire process.

The farmer's part is merely tending the soil, while the industry decides on questions of sowing and harvesting, furnishes seed and takes care of disease-control. The same goes for alfalfa for fodder pellets which is now grown on many holdings that have no livestock of their own. Sugar-beet has been a regional crop for almost a century, and more recently we have seen pig breeding centres, "egg-factories" and broiler-production, "duck-producers", tomato-gardeners, and fruit-growers. There is also a tendency to cultivate marginal areas in the west more extensively — enlarging the grass-areas for livestock in contrast to the cereal fields in the eastern parts.

Specialisation in Danish agriculture will hardly proceed speedily without the direction of superior planning authorities, which do not exist in Denmark at present. An investigation similar to that presented here may perhaps be instrumental in furthering the spread of spontaneous specialisation, merely by offering the very proof of its existence.

As is generally known a healthy agricultural environment rests on the interaction between cultivated land and nature, or rather nature's constantly disturbed states of equilibrium. Where little or no wild-life is left, the presence of which is a check on crop parasites, the balance is upset. Specialisation will disturb the balance of nature even more than the mixed farming of today.

So that seed-crops and fruit-trees may be pollinated, it is essential to secure the conditions of life of bees. Mono-culture cropping is only possible in areas where crops are grown that either do not attract bees, or if there are bees present that flower only for a short period. In this way the harm that can be caused by insecticides can be avoided. Thanks to hormone-treatment fields are nearly devoid of flowering weeds. Field edges are cut, meadows are drained and converted into fields, but bees claim untilled areas with wild flowers.

A future challenge when considering structural changes in Danish agriculture will be how to continue to increase the output from arable land with a steadily dwindling labour-force with due regard to geographical differences. Economic growth will necessarily be connected with increasing industrialisation. But in the long view energy supply problems cannot be disregarded, problems that may also have a serious effect on agricultural profitability.

But whatever the future will bring the farmer will always have the privilege of being in an occupation which has not been dehumanised by technology. Contrary to the fate of many other people there is a quality of life not only in his leisure time but during working hours as well.

APPENDIX

G. GREENLAND AND THE FAROES

Outside Denmark proper lie Greenland and the Faroes as parts of the Danish Kingdom. Greenland, the largest island in the world, has a total area of 2,175,000 sq.km, of which 1,833,900s q.km are under ice. The southernmost point is situated as far south as 59° 46′ N. lat. Of the ice-free area, 341,700 sq.km, some 150,000 sq.km, chiefly in the south-western part, may be classed as inhabited. There are 48,000 inhabitants (1971), of which 40,000 were born in Greenland.

In 1953 the administrative status of colony was abolished and Greenland is now part of the Kingdom.

The Faroe Islands *(Föroyar)* are situated in the Atlantic between 6° 15′ and 7° 41′ W. long. The latitude, between 61° 26′ and 62° 24′ N. lat., is six degrees further north than Köbenhavn. The Faroes comprise 1400 sq.km and have 39,000 inhabitants (1973).

Since 1948 the Faroe Islands in most respects have been a self-governing region of the Kingdom.

AGRICULTURE IN GREENLAND

A rise in sea temperature between 1920 and 1935 brought about a change in marine fauna, cod now being common near Greenland.

These new possibilities coincided with a large increase in population and led to a complete reorganisation of the Greenlanders' way of life. Seal hunting, previously the main pursuit, had provided them with virtually all the requirements of a subsistence economy: food, garments, kayaks, oomiaks, lighting and heating. But as fish, now the main trade, can provide only part of the food supply, people have been concentrating in the trading stations where the fish can be sold. Seal hunting is no longer important except in the northernmost habitations.

Part of the new economic developments necessitated by the decline of the seal and the increase in population has taken place within agriculture. Greenland is of course a marginal region as far as agriculture is concerned and it is debatable whether the limit of profitable agriculture under sub-arctic circumstances has been exceeded.

Though the longest frost-free periods occur in the northern continental districts of Greenland outdoor cultivation is of course only possible towards the south where permafrost is absent (*Fig. 33* and *Photo 13*). The outer coasts here are often foggy with high precipitation while the weather in the innermost parts of the long fjords near the inland ice is drier and more continental with more hours of sunshine and relatively high summer temperatures. Even there the growing season is very short, but July and August are frost-free and the accumulated temperature is relatively high because of the length of day.

As a consequence of the slow rate of weathering the surface has an extremely thin layer of soil for the natural vegetation. The parent material consists mainly of morainic deposits with plenty of stones, the small particles being to a large degree washed out. Tillable soil is found almost entirely in the same places which were used by the Norsemen about 1000 years ago.

The soil is difficult to work, and it is easy for water to penetrate the thin, sandy layers, so that irrigation may be profitable in some places. Some potatoes, roots and various vegetables can be grown. Under favourable conditions the yield of potatoes may be about the average for Jylland. By the use of hothouses, heated with Greenland coal, the production of vegetables could be considerably advanced, the summer being hereby prolonged by 2—3 months. Hothouses are found as far north as Upernavik (73° N).

The early garden turnip, rich in vitamin C is, like the potato, grown for local consumption. Experimental cultivation of different vegetables is being carried out near Julianehåb.

Grain will not ripen but can profitably be harvested green and dried for hay.

The fairly rich natural vegetation consists of shrubs, mosses, lichens, grasses and herbs, which can be used for grazing and hay. The yield of hay may be increased by the use of fertilisers and irrigation and by establishing fields for growing grass crops after stone clearing and soil preparation. These conditions have made possible the development of sheep farming as the most important branch of agriculture in Greenland.

Summer grazing is good and abundant, and if the mountains are clear of snow half the optimum need of food is also secured during winter time, but far less than half of the necessary protein. Sheep breeding is consequently especially vulnerable as regards wintering, but the losses could be kept to a minimum if there were always sufficient winter fodder for disposal.

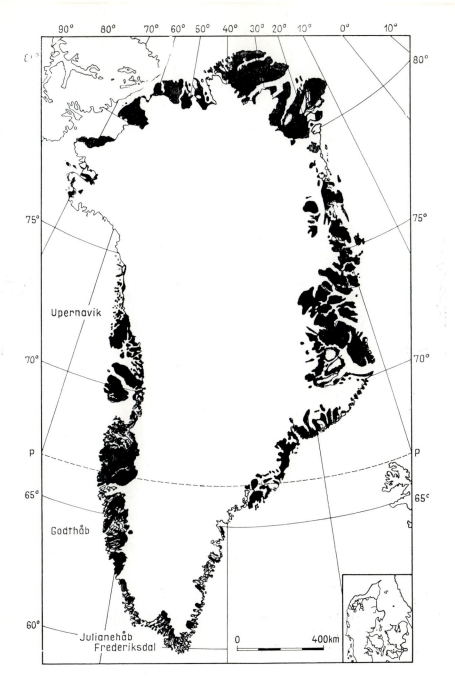

Fig. 33. Greenland. Black = ice-free areas

Photo 13. Sheep are gathered before slaughter, K'agsiarssuk, Greenland (by Kampp)

The increase of the tilled areas may consequently be considered the most important problem for the sheep farmer.

As transport costs make it impossible to import fodder, sea-weed is used in dehydrated condition as stable fodder, while the sheep may graze upon it themselves. Also fish is greatly used, fishmeal being from the waste products of the fillet factories.

In the Julianehåb district sheep farming is partly run as a side-line near the outer coasts and partly as principal occupation in the inner parts of the fjords. In 1971 there were 34,500 sheep in Greenland.

As a Greenland sheep yields only 2 kg wool and often loses part of it in the mountains, it is kept mainly for its meat. Every year some 15,000 sheep and lambs are slaughtered.

Modern sheep farming was started in 1906 in Frederiksdal with Faroese sheep; it was supported by the administration which in 1915 founded a sheep breeding station and an agricultural college for the education of Greenland sheep breeders. The state supports new sheep breeders with establishment loans.

AGRICULTURE IN THE FAROES

"Seyda ull er Föroya gull", the wool of the sheep is the gold of the Faroes, is a saying from the time when agriculture was the chief occupation in the Faroe islands. Even in 1801 over 80 per cent of the population subsisted on agriculture, whereas now barely five per cent of the working public is solely occupied in farming, the production of which is very limited.

About six per cent of the area is enclosed, and for the most part under cultivation. The remaining 94 per cent is outfield.

The first farms were established by the early settlers around 900 A. D. During the Roman Catholic period more than half of the area was taken over by the Church but after the Reformation it became Crown property. Although these copyhold farms could only be leased, a resident family enjoyed the right of inheritance, the property being inherited undivided. This state of affairs was also maintained after the 400 estates were taken over by the Faroese authorities in 1955.

Like the "royal estates" the freehold land consists of land-registered enclosed properties with rights of pasturage, bird-catching etc. in the outfield. The number of freehold properties varies greatly, as they can be partitioned on the death of an owner. In this way the freehold land has come into so many hands that one can hardly talk about a proper farm any more, as there is no limit to how small a plot might become when divided equally between a number of heirs.

A third category of holding is the socalled *tröd*, "croft", a section of land from the outfield which has been enclosed and turned into arable land. They number about 1000, and are usually some 12,000 square meters in size, which only enables the tröd-holder to keep a cow and to supply his family with potatoes and vegetables. He enjoys no rights in the outfield.

The enclosed areas of the islands are found on the slopes or in hollows mostly as narrow, often winding strips. The effective vegetation period is brief, and climate, soil, and the conditions of the terrain severely limit the choice of plants to be cultivated. More than 90 per cent of the enclosed areas are laid down for hay and grazing. Formerly the spade was the most important implement but during recent years agricultural machinery has become increasingly common.

Crop rotation was earlier applied to the growing of barley, the grassland being laid down for grain every seventh year and heavily manured (*Fig. 34*). However, the cultivation of cereals gradually came to an end as fishery became an all-year-round occupation, and the population increasingly deserted a subsistence economy in favour of a monetary one (*Photo 14*).

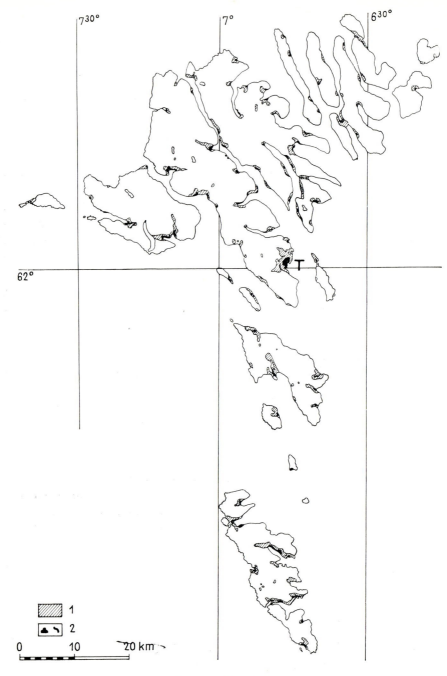

Fig. 34. The Faroes

1 = enclosed land; 2 = towns and villages (*bygder*); T = Tórshavn, the Capital

Photo 14. Vestmanna, the Faroes. Behind the village are the enclosed fields, and on the mountain the outfield. In the foreground hay is being dried (by Kampp)

Today, apart from some gardening, practically nothing is grown beyond grass and potatoes; during World War II the islands were self-supporting in potatoes.

Until about 15 years ago the horned cattle population was about 4000, a good 3000 of which comprised dairy cows; the figures have now dropped to 2500 and 1900 respectively, contemporaneously with an increase in milk production. Thanks to more modern cowsheds and some degree of mechanised operation a few farms have been able to increase the size of their herds. During the summer the cattle are grazed in those parts of the outfield bordering on the enclosed areas; in the winter they are usually kept in cowsheds and fed on silage, hay and imported concentrates.

Horses were formerly important for carrying purposes although they were never used much as draught animals in agriculture. The number of horses was never great and has now declined to about 300 because of the increasing use of cars and lorries.

Poultry-keeping has increased particularly since the introduction of poultry farms.

The outfields are uncultivated and serve as pasture for sheep nearly all year round, and for horses and cows during the summer. Sheep breeding is the most important branch of agriculture. About 70,000 are kept, a figure, which has remained virtually unchanged as fewer sheep would not completely utilise the grazing, whilst a greater number would give rise to overgrazing with the danger of soil erosion.

Most of the sheep are jointly owned by the farmers in the enclosed areas with outfield rights. The returns from both wool and slaughtering are shared between them in proportion to the size of their enclosed holdings. The sheep live, though under a certain supervision, in a semi-wild state for the greater part of the year, but during the winter they are moved to the lower districts of both the enclosed and outlying areas. Twice a year they are collected and driven into folds; in the spring for gathering their wool. This is usually so loose that it can be pulled off; otherwise it is cropped or shorn. In the autumn the selection of animals for slaughter takes place and about 40,000 sheep and lambs are slaughtered annually. The meat is salted or wind-dried, and is entirely consumed in the Faroes. Despite this it is necessary to import large quantities of frozen mutton to satisfy local demand.

Although originally it was not the general practice to give the animals any fodder supplement during the winter, it is now becoming increasingly common to build folds in the outlying regions and to feed the animals at times when they have difficulty in fending for themselves.

The improved earning possibilities of the fisheries have particularly affected the small farms and a combination of fishing and a small plot of land with one or two cows is now not as common as was once the case. On the whole the importance of agriculture to the Faroes today is comparatively modest, but it could presumably be augmented by further mechanisation and a greater stability with regard to ownership of land. There is a trend in agricultural circles to consolidate the plots into farms of such a size that with modern equipment can support a family.

REFERENCES

Begtrup, G. (1803 – 1812): *Beskrivelse af Agerdyrkningens Tilstand i Danmark*. Denmark.

Dalbro, S. (1967): Erhvervsfrugtavl og klima i Danmark. *Erhvervsfrugtavleren*, 10.

Det landökonomiske Driftsbureau: Landbrugets prisforhold 1972 – 1973.

Frandsen, K. E. (1969): Lidt om landbrugsgeografi og naturlandskab. *DLH*.

Gissel, S. (1968): Die Dreifelderwirtschaft auf Seeland bis 1700. *Geogr. Zeitschrift.*

Hansen, V. (1943): Östjylland, en geografisk provins. *Geografisk Tidsskrift.*

Hansen, C. Riise and Steensberg, A. (1951): Jordfordeling og udskiftning. *Videnskaberns Selskab*, Köbenhavn.

Humlum, J. and Skjödt, A. (1973): Europe's production of and trade in fruit and vegetables. *Kulturgeografi* 120.

Kampp, Aa. H. (1936): *Indledende undersøgelser over de geografiske betingelser for erhvervsfrugtavl i Danmark*. Helsinki.

Kampp, Aa. H. (1939): Forarbejder til et dansk landbrugsatlas. *Tidsskrift for Landökonomi.*

Kampp, Aa. H. (1942): Dansk frugt og verdensmarkedet. *Tidsskrift for Landökonomi.*

Kampp, Aa. H. (1943): Formindskes vort landbrugsareal? *Tidsskrift for Landökonomi.*

Kampp, Aa. H. (1944a): Den danske udstykning i geografisk belysning. *Naturhistorisk Tidende.*

Kampp, Aa. H. (1944b): En metode til inddeling i landbrugsgeografiske områder. *Geografisk Tidsskrift.*

Kampp, Aa. H. (1952): Dansk landbrugsgeografi og statistik. *Geografisk Tidsskrift.*

Kampp, Aa. H. (1956a): Die landwirtschaftliche Regionen Dänemarks. *Geographische Rundschau.*

Kampp, Aa. H. (1956b): Die dänische Agrarreform im 20. Jahrhundert. *Geographische Rundschau.*

Kampp, Aa. H. (1957): Om danske markfröafgröders geografiske placering. *Tidsskrift for fröavl.*

Kampp, Aa. H. (1958): *Les régions agricoles du Danemark*. Norois, Lille.

Kampp, Aa. H. (1959a): Some agro-geographical investigations of Denmark. *Kulturgeografiske skrifter*, 6.

Kampp, Aa. H. (1959b): Some types of farming in Denmark. *Oriental Geographer.*

Kampp, Aa. H. (1959c): Utilization of arable land on moraine landscape and on outwash plains in Denmark. *Geografisk Tidsskrift.*

Kampp, Aa. H. (1960): Danish agricultural subdivision and the Majorats. *Geografisk Tidsskrift.*

Kampp, Aa. H. (1962): The agricultural geography of Mön. *Erdkunde*, Bonn.

Kampp, Aa. H. (1963): Die heutige Lage der Landwirtschaft in Dänemark. *Geographische Rundschau.*

Kampp, Aa. H. (1964a): Die Aufteilung der dänischen Majorate. *Geographische Rundschau.*

Kampp, Aa. H. (1964b): *De levende råstoffers geografi.* Köbenhavn.

Kampp, Aa. H. (1965a): Landbrugsregioner i Danmark, deres udnyttelse og udvikling. *Byplan.*

Kampp, Aa. H. (1965b): *Landbrugslandskabet.* In K. and Aagesen: Det danske Kulturlandskab. Köbenhavn.

Kampp, Aa. H. (1967): Fåreavl i Grönland og i nogle andre nordatlantiske områder. *Grönland.*

Kampp, Aa. H. (1969a): Fruit growing and climate in Denmark. *Landbrugsgeografiske småskrifter, DLH.*

Kampp, Aa. H. (1969b): *Denmark.* World Atlas of Agriculture. Verona.

Kampp. Aa. H. (1970a): Erhvervsgeografisk belysning af dansk landbrug. *LOK-studiebögerne.*

Kampp, Aa. H. (1970b): Landbrugsgeografisk materialesamling. *DLH.*

Kampp, Aa. H. (1970c): The changing patterns of land use and the agro-geographical division of Denmark. *Geografia Polonia,* 19.

Kampp, Aa. H. (1970d): Agerlandets geografi. *Danmarks Natur,* 8.

Kampp, Aa. H. (1970e): Entwicklungstendenzen der dänischen Landwirtschaft. *Wissensch. Abhandl. Deutsch. Geographentag,* Kiel.

Kampp, Aa. H. (1972a): Changes in the distribution of agricultural land in Denmark. *Geografisk Tidsskrift.*

Kampp, Aa. H. (1972b): Changing patterns of agricultural settlements in Denmark. *IGU Symposium,* Pécs, Hungary.

Kampp, Aa. H. (1973a): Tendencies in the development of Danish agriculture. Agricultural Typology and Land Use. *IGU,* Hamilton.

Kampp, Aa. H. (1973b): *Färöerne.* Köbenhavn.

Kampp, Aa. H. and Frandsen, K. E. (1967): A farm in the village. *Geografisk Tidsskrift.*

Kampp, Aa. H. and Frandsen, K. E. (1970): Lidt om trevangsbrug i Europa. *Naturens Verden.*

Kristofferson, A. (1931): Regionalgeografiska Studier i mellersta Jylland. *Svensk Geografisk Årsbok.*

Rigsombudsmanden (1973): *Årsberetning 1972.* Tórshavn.

Schou, A. (1949): The landscape. *Atlas of Denmark I.* Köbenhavn.

Schon, A. *et al.* (1967—71): *Danmarks Natur I—XII.*

Smith, C. T. (1967): *An historical geography of western Europe before 1800.* London.

Statens Forsögsvirksomhed i plantekultur (1942): 326. meddelelse. a. Forsög med rodfrugter 1925—36. Köbenhavn.

Statistical Yearbook 1973. Köbenhavn.

Statistiske undersögelser 22. Landbrugsstatistik 1900—65. Köbenhavn, 1968.

Statistiske undersögelser 25. Landbrugsstatistik 1900—65, II; 1969—1972.

Statistiske meddelelser 1972, 9. Landbrugsstatistik *1971.*

Sömme, A. (1960): *A geography of Norden.* Oslo.

Thomsen, C. (1972): Landbruget 1971. *Tidsskrift for Landökonomi,* 1.

Thomsen, C. (1973): Landbruget 1972. *Tidsskrift for Landökonomi,* 1.

Thomsen, C. (1974): Landbruget 1973. *Tidsskrift for Landökonomi,* 1.

Vahl, M. (1942): De geografiske provinser i Danmark og nogle forhold inden for disses rammer. *Svensk Geografisk Årsbok.*

RESEARCH INSTITUTE OF GEOGRAPHY
HUNGARIAN ACADEMY OF SCIENCES

GEOGRAPHY
OF WORLD AGRICULTURE
5

György Enyedi, Editor-in-Chief (Hungary)

AKADÉMIAI KIADÓ · BUDAPEST 1975

Aa. H. KAMPP

AN AGRICULTURAL GEOGRAPHY
OF DENMARK